Let's Learn About Farm Machines

HARVESTING EQUIPMENT

Therese M. Shea

Enslow Publishing
101 W. 23rd Street
Suite 240
New York, NY 10011
USA
enslow.com

Words to Know

blade A flat, sharp part of a weapon or tool that is used for cutting.

conveyor A wide strip of material that carries things from one place to another.

elevator A machine that raises and lowers objects.

engineer A person who plans and builds machines, buildings, products, and ways of doing things.

grain A plant that makes a seed used for food.

harvesting Gathering crops.

invent To create something useful for the first time.

tractor A large machine that has two large back wheels and two smaller front wheels. It pulls farm equipment.

CONTENTS

Crops are harvested in different ways. Strawberries are ready to eat, but grains like wheat take more work.

Farm Crops

Crops are plants that are grown by farmers. Some are picked and eaten, like strawberries. Others are picked and separated into parts. Wheat and other grains need to be separated into parts.

FAST FACT

We use wheat seeds called kernels to make bread and other foods.

All of these potatoes need to be harvested. Machines make the job much quicker and easier.

Huge Harvests

Some farms are huge. Fields of crops seem to go on forever. Picking a crop by hand can take a long time. Luckily, many farmers have harvesting machines to help!

FAST FACT

People once used sharp tools called scythes to chop down tall crops.

The McCormick reaper from 1831 cut crops in the field. Workers then had to gather and bundle the crops.

Early Harvester

The first modern harvesting machine was **invented** in the early 1800s. It was called a reaper. It was pulled by horses and cut wheat.

FAST FACT

Early reaper machines left the cut crops in the field.

The combine harvester was able to do more work than the reaper.

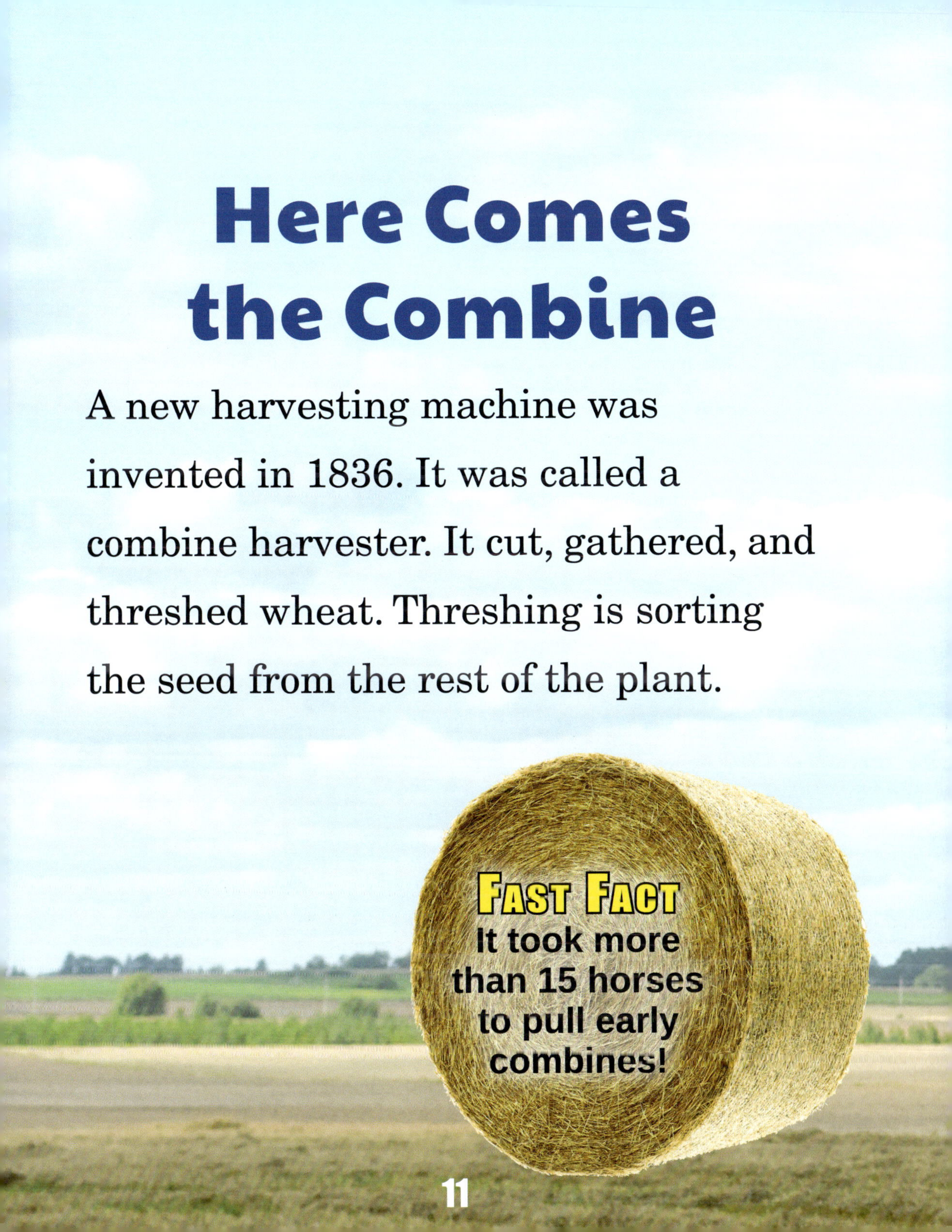

Here Comes the Combine

A new harvesting machine was invented in 1836. It was called a combine harvester. It cut, gathered, and threshed wheat. Threshing is sorting the seed from the rest of the plant.

FAST FACT

It took more than 15 horses to pull early combines!

Horses and tractors were no longer needed when self-propelled combines were invented.

On the Move

Horses pulled early combine harvesters, or combines. Then, **tractors** pulled combines. Today, combines can be driven without tractors. They are called self-propelled combines.

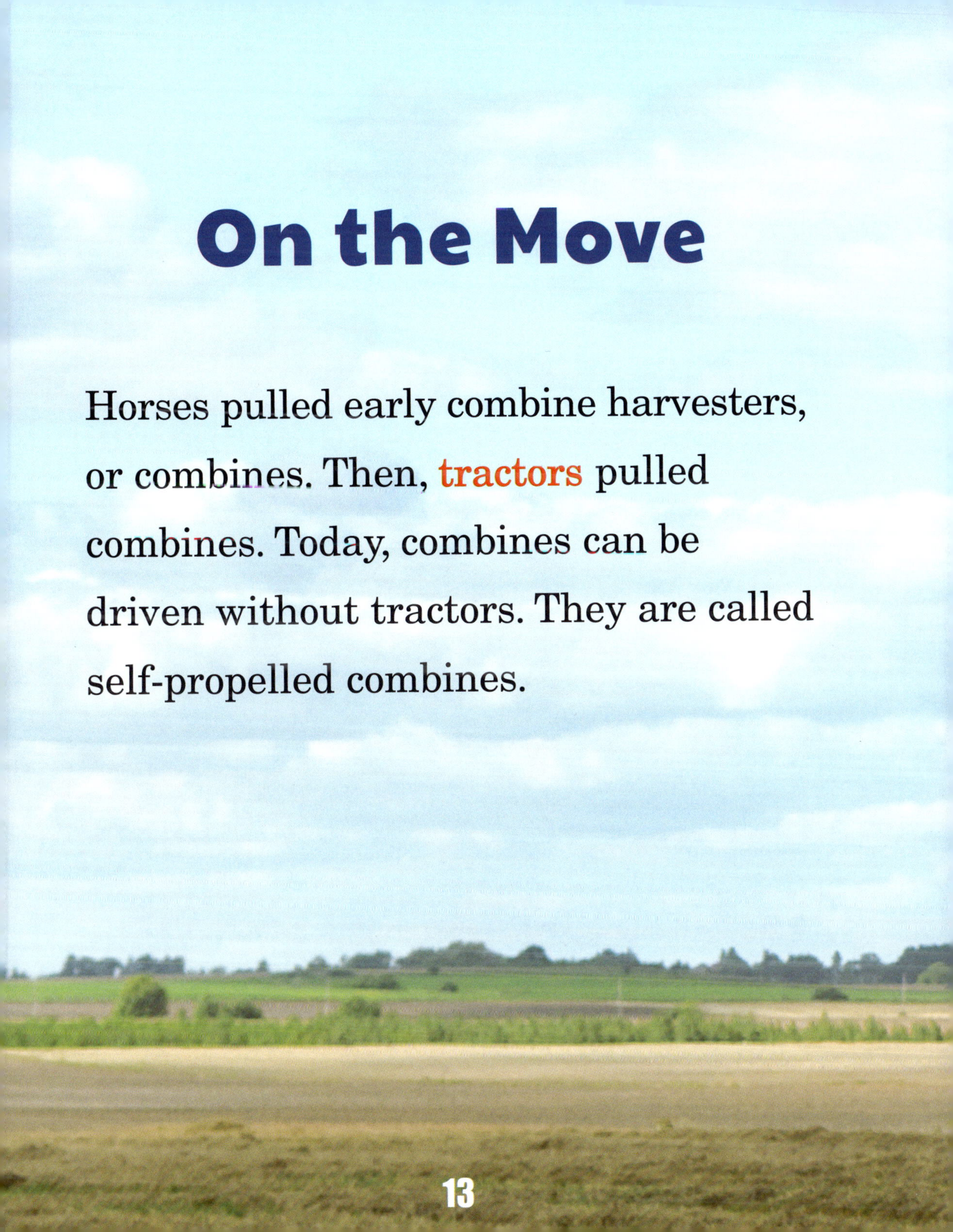

A modern combine harvester cuts, separates, and cleans wheat.

Many Jobs

Today's combines do a number of jobs. They cut crops. They feed them into a machine. They separate the parts. They gather and clean the parts we use.

FAST FACT

Combines have **blades** to cut crops, **conveyors** to move them, and a drum to thresh them.

The header on this combine is made for harvesting carrots.

All Kinds of Crops

Engineers invented combines for many kinds of crops. Combines harvest corn, cotton, rice, tomatoes, carrots, and potatoes. They use different headers, or front pieces, for different crops.

The combine sends the crops it has collected into a nearby truck.

Up the Elevator

A combine collects the crop in a tank. When the tank is full, an elevator in the combine pushes the crop out. A trailer next to the combine collects the crops.

FAST FACT

Unneeded plant parts are called chaff. They are pushed out the back of the combine.

Modern combines make harvesting crops fast and easy.

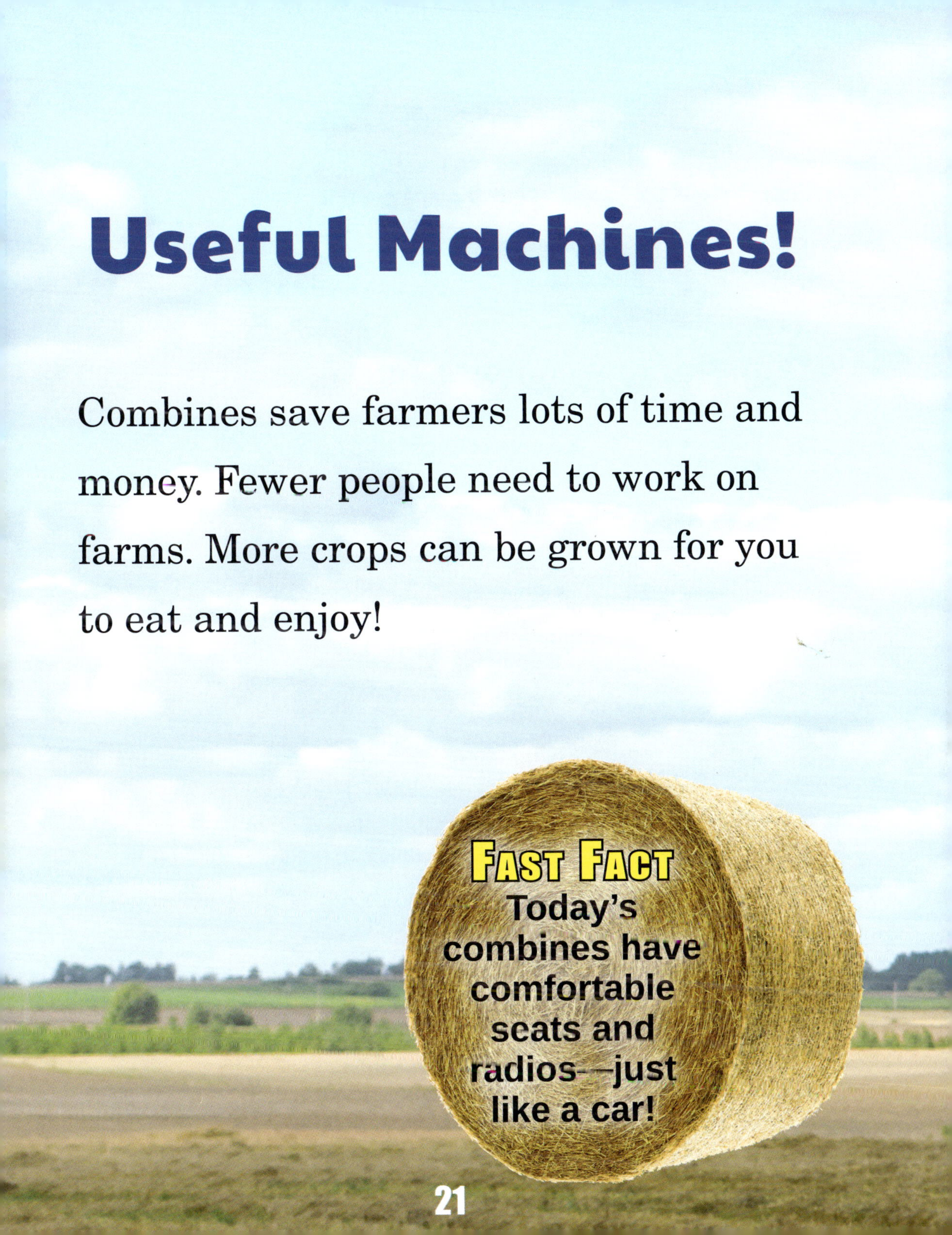

Useful Machines!

Combines save farmers lots of time and money. Fewer people need to work on farms. More crops can be grown for you to eat and enjoy!

FAST FACT

Today's combines have comfortable seats and radios—just like a car!

Activity

Find That Combine!

What kind of combine harvests your favorite foods?

YOU WILL NEED:

- Large poster board
- Food store ads
- Markers
- Scissors
- Glue

Step 1: Think of your four favorite farm crops, such as corn, potatoes, rice, or oats.

Step 2: Write the name of each crop in a corner of your poster board. Near each name, draw a picture of the crop or glue on a picture of it.

Step 3: Ask an adult to help you use the internet. Find a picture of the combine that harvests each crop.

Step 4: Print a picture of each combine. Glue it next to the crop on your poster.

Step 5: Give your poster a fun title.

A combine is used to harvest grapes.

Learn More

Books

Dittmer, Lori. *Harvesters*. Mankato, MN: Creative Education, 2018.

Maimone, S. M. *Harvesters*. New York, NY: Gareth Stevens, 2017.

Wilson, Hannah. *Combine Harvesters*. New York, NY: Kingfisher, 2015

Websites

Combine Harvesters
www.explainthatstuff.com / howcombineharvesterswork.html
Read more about what happens inside a combine.

Combines
historylink101.com / lessons / farm-city / combine.htm
Read a short history of the combine.

Index

Published in 2020 by Enslow Publishing, LLC.
101 W. 23rd Street, Suite 240, New York, NY 10011

Library of Congress Cataloging-in-Publication Data

Names: Shea, Therese M., author. |
Title: Harvesting equipment / Therese M. Shea.
Description: New York : Enslow Publishing, 2020. | Series: Let's learn about farm machines | Includes bibliographical references and index. | Audience: Grades K-3.
Identifiers: LCCN 2019006123| ISBN 9781978513112 (library bound) | ISBN 9781978513099 (pbk.) | ISBN 9781978513105 (6 pack)
Subjects: LCSH: Harvesting machinery--Juvenile literature. Harvesting--Juvenile literature.
Classification: LCC TJ1485 .S54 2020 | DDC 631.5/50284--dc23
LC record available at https://lccn.loc.gov/2019006123
Printed in the United States of America

Photos Credits: Cover, p. 1 Paul R. Jones/Shutterstock.com; interior pages background landscape Jurgis Mankauskas/Shutterstock.com; p. 4 Mirko Graul/Shutterstock.com (strawberries); Kostiantyn/Shutterstock.com (wheat); pp. 5, 7, 9, 11, 13, 15, 17, 19, 22 (hay bale) photomaster/Shutterstock.com; p. 6 oticki/Shutterstock.com; p. 8 Print Collector/Hulton Archive/Getty Images; p. 10 Universal History Archive/Universal Images Group/Getty Images; p. 12 Alfred Eisenstaedt/The LIFE Picture Collection/Getty Images; p. 14 Bloomberg/Getty Images; p. 16 Konggulerod/Wikimedia Commons/File:Asa_Lift_6_row_top_lifter_carrot_harvester/CC BY-SA 3.0; p. 18 TTstudio/Shutterstock.com; p. 20 Deyana Robova/Alamy Stock Photo; p. 23 Maleo/Shutterstock.com; cover, p. 1 logo element (tractor) Krivosheev Vitaly/Shutterstock.com.